Light Bulb

BY KATHLEEN WEIDNER ZOEHFELD • ILLUSTRATED BY STEPHANIE DEHENNIN

KANEPRESS

AN IMPRINT OF BOYDS MILLS & KANE

New York

For Geoff —KWZ

To all the curious kids, the tinkerers, and the ponderers. They are the future generation of scientists, artists, and engineers, working to better our world. —SD

Special thanks to Paul Israel, Director & General Editor of the "Thomas A. Edison Papers" at Rutgers University

For information about permission to reproduce selections form this book, please contact permissions@bmkbooks.com.

Kane Press
An imprint of Boyds Mills & Kane, a division of Astra Publishing House
kanepress.com
Printed in China

Library of Congress Cataloging-in-Publication Data
Names: Zoehfeld, Kathleen Weidner, author. | Dehennin, Stephanie, illustrator.
Title: Light bulb / by Kathleen Weidner Zoehfeld ; illustrated by Stephanie Dehennin.
Description: First edition. | New York : Kane Press, an imprint of Boyds Mills & Kane, [2021] | Series: Eureka! | Audience: Ages 4-8. | Audience: Grades K-1. | Summary: "A nonfiction 'biography' of the light bulb from the electric precursors like the arc lamp to Thomas Edison's invention and beyond to current innovations in light bulb technology"—Provided by publisher.
Identifiers: LCCN 2020046035 (print) | LCCN 2020046036 (ebook) | ISBN 9781635923964 (paperback) | ISBN 9781635923957 (hardcover) | ISBN 9781635924763 (ebook)
Subjects: LCSH: Light bulbs—History—Juvenile literature. | Lighting—History—Juvenile literature.
Classification: LCC TK4351 .Z64 2021 (print) | LCC TK4351 (ebook) | DDC 621.32—dc23
LC record available at https://lccn.loc.gov/2020046035
LC ebook record available at https://lccn.loc.gov/2020046036

10 9 8 7 6 5 4 3 2 1

IMAGINE A WORLD WITHOUT ELECTRIC LIGHTS.

   Your house has no lamps and no lights overhead. You don't even have a flashlight! When the sun goes down, everything gets dark. There's nothing to do but crawl under the covers and wait until sunrise, right?

   Well, not exactly!

If you grew up in the 1800s, like the inventor Thomas Alva Edison, you might light a candle or an oil lamp. You might even have gaslights in your house. But gaslights were expensive. And candles? A breeze could blow them out.

When he got older, Edison loved to work late into the night. He'd write down ideas for new inventions and sketch in his notebooks. Sunset didn't stop him. He lit an oil lamp and kept going.

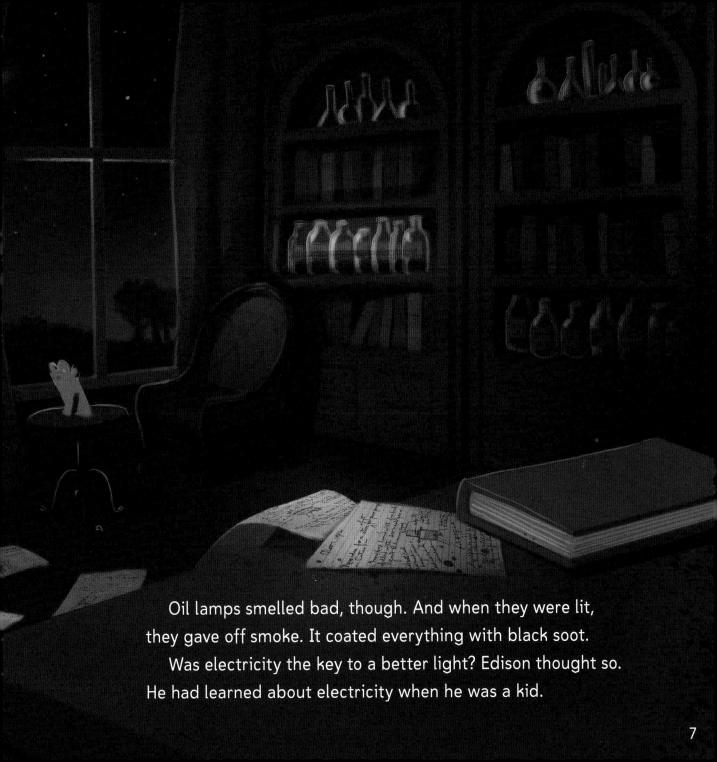

Oil lamps smelled bad, though. And when they were lit,
they gave off smoke. It coated everything with black soot.

Was electricity the key to a better light? Edison thought so.
He had learned about electricity when he was a kid.

In the early 1800s, a scientist named Humphry Davy had created a type of electric light. He called it an arc lamp. Electric current jumped across a small space between two carbon rods, making a flash like lightning. But Davy's lamps burned out very quickly.

Later scientists made arc lights that lasted longer. Most were used for lighting streets at night. Indoors, they were annoying, and sometimes dangerous. They buzzed loudly and threw off sparks that could start fires. And their light was way too bright.

A softer, safer, quieter light that could be used in homes? Other inventors had tried. They had failed. Could Edison be the one to succeed?

In 1878, Edison built a big laboratory in Menlo Park, New Jersey. He told his crew that their main job was to come up with new ideas. The lab was nicknamed "The Invention Factory."

And new ideas poured out of it! Soon, everyone began calling Edison the "Wizard of Menlo Park."

They improved the sound of Alexander Graham Bell's new invention, the telephone.

They invented the world's first phonograph. It could record and play back sounds.

They came up with ways to make better telegraphs. Telegraphs sent coded messages over great distances.

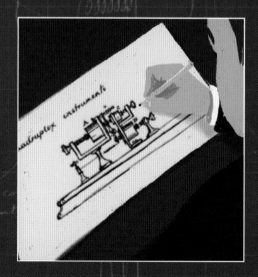

Edison put electric light on his list of projects. But the problems with it were difficult to solve. By 1878, he had so many things going on, he needed a break!

That summer, a couple of friends suggested a vacation.

Edison came home full of new ideas. Just one day after he returned, he drew three sketches in his notebook. He labeled them "Electric Light."

Edison told newspaper reporters all about his idea. Soon he'd show the world his new light!

But before long, Edison realized it didn't work. And now everyone was watching. Everyone expected great things.

The pressure was on!

Edison told his team not to worry. Failure was good, he said. It was a way of finding out what didn't work. Eventually they'd find out what *did* work.

They kept trying. Most of their early light bulb designs were very costly to make. All burned out too quickly.

The glowing part—which Edison named the filament—was the problem. Edison knew they needed a special material for their filament. Something that was cheap. Something that would last. But what?

••• HOW AN ELECTRIC LIGHT BULB WORKS •••

Electric current starts at a power source, such as a battery. It flows through a wire to a filament.
The filament is mounted between two wires inside a glass bulb.

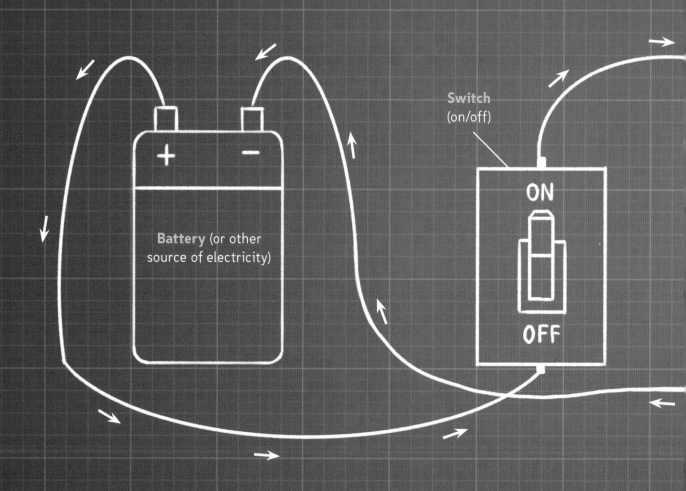

Switch
(on/off)

ON

OFF

Battery (or other
source of electricity)

The electric current passes through the filament from the first wire to the second wire. The filament begins to heat up and glow. In this way, the energy of electricity is changed into heat and light.

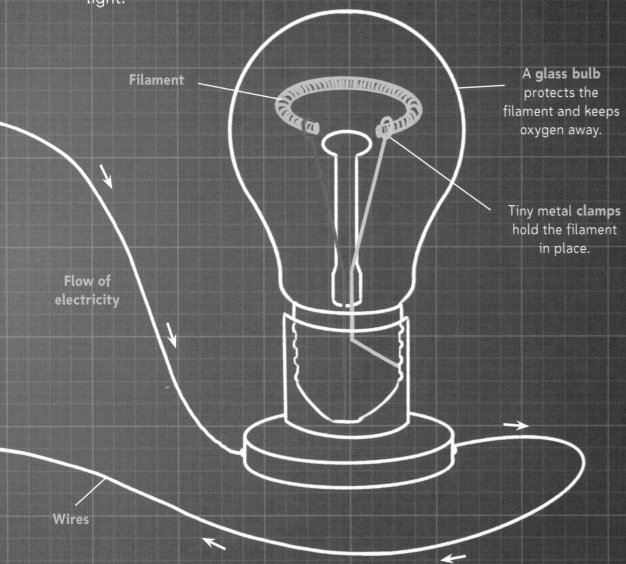

Filament

A **glass bulb** protects the filament and keeps oxygen away.

Tiny metal **clamps** hold the filament in place.

Flow of electricity

Wires

Months went by with no luck.

Then one night in early October 1879, Edison was fiddling with some black carbon soot. It had coated the glass globe of one of the lab's oil lamps. He rolled the sticky soot between his fingers into a thin thread. Could pure carbon be the answer?

Maybe! If only it wasn't so crumbly!

Carbon is one of the main elements in all living things, such as animals and plants. Wood, paper, and cotton are examples of common items that are made from plants, so they are high in carbon.

Edison baked, or carbonized, a bit of cotton thread and used it as a filament. He and his crew could hardly believe their eyes. The filament glowed for over fourteen hours! They were on the right track!

But could they do better? The team tested dozens of different carbonized filaments—wood, paper, coconut shell, fishing line, even human hair! The winner was a thick paper called Bristol board. It lasted more than 100 hours.

Have you ever burned a piece of toast? That black coating is carbon. When a material made up of mostly carbon is baked at a high temperature, it turns black. All the other elements in it have been burned away. This process is called carbonization.

It was time to celebrate! Edison set up a dazzling display of lamps in Menlo Park. On New Year's Eve 1879, he invited everyone to see.

Thousands of people followed the lights from the train station to the lab.

What a soft and steady light! A light without flame! Everyone knew this was an important moment in history.

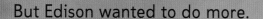

But Edison wanted to do more.

He and his crew set up a big generator in New York City. They ran wires underground. They put light fixtures in homes and offices. People plugged in their Edison lamps.

On September 4, 1882, at three o'clock in the afternoon, Edison turned on the generator and lit up an entire square mile of New York City.

Soon everyone wanted electric lights, too!

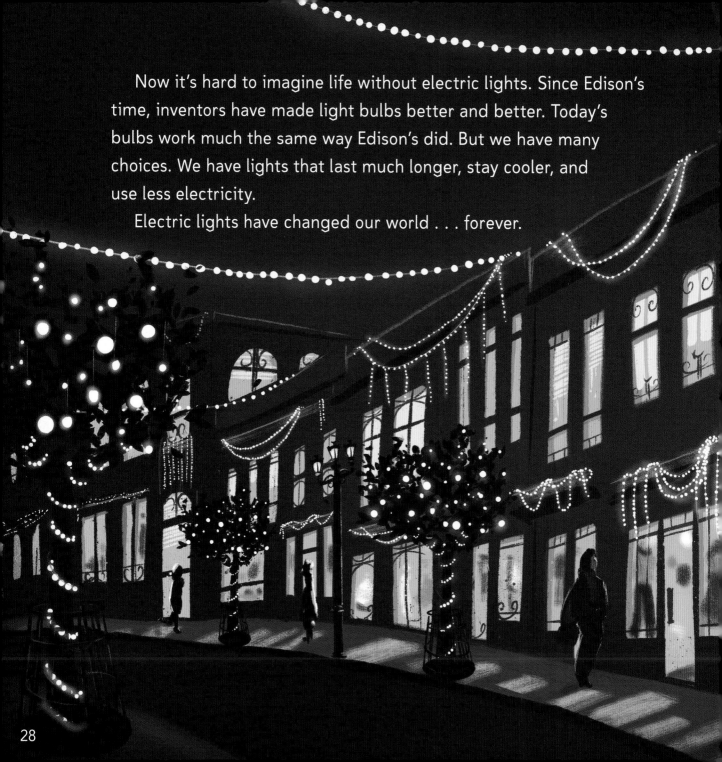

Now it's hard to imagine life without electric lights. Since Edison's time, inventors have made light bulbs better and better. Today's bulbs work much the same way Edison's did. But we have many choices. We have lights that last much longer, stay cooler, and use less electricity.

Electric lights have changed our world . . . forever.

••• FACTS ABOUT LIGHTS •••

• In 1880, Edison's team got a light bulb to glow for more than 1,000 hours. The filament was made of carbonized bamboo.

• The British inventor Joseph Swan made a light bulb around the same time as Edison. Swan and Edison respected each other's work. They even formed a new company together in 1883.

• Around 1911, scientists found that filaments made of a special metal, called tungsten, lasted much longer than carbon.

- Curvy, colorful neon lights became popular for signs and advertising in the 1920s.

- The first fluorescent tube lights were made in the 1930s. They were used in many offices and schools.

- Today, red, yellow, and green LED lights are used for traffic signals and to light up the numbers on digital clocks.

- Soft white, energy-saving LED light uses a combination of blue, red, and green LEDs. The inventors won the Nobel Prize for their work!

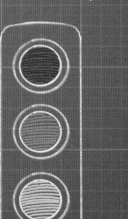

LIGHTING THE FUTURE

Inventors are always looking for ways to improve things. Edison gave the world a clean, safe form of light, without smoke or flame. Today the challenge is to make lights that use less energy.

When it's turned on, an Edison-style bulb gets very warm. Modern LED lights stay cool. Warm lights waste energy in the form of heat. Cool lights save energy.

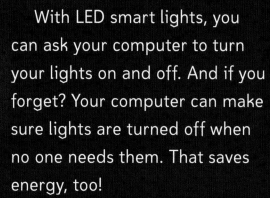

With LED smart lights, you can ask your computer to turn your lights on and off. And if you forget? Your computer can make sure lights are turned off when no one needs them. That saves energy, too!